THE FARM BOOK

text by   CHARLES E. ROTH
R. JOSEPH FROEHLICH

illustrations by   RUSSELL BUZZELL

THE
MASSACHUSETTS AUDUBON
SOCIETY
LINCOLN • MASSACHUSETTS

This edition published by
HARPER & ROW, PUBLISHERS
New York, Hagerstown,
San Francisco, London

FOR

All those fellow travelers on this planet whose lives are inextricably linked to the products of the land and the farm people who work to share the bounty of the land with all of us.

ACKNOWLEDGMENTS

The authors of this volume want to express their deep appreciation to the Board of Directors of the Massachusetts Audubon Society for their faith in and support of this project; to the M.A.S. staff for much assistance of many kinds; and to Ray Broekel for his generous guidance and technical assistance along the way.

Our special thanks to the Massachusetts Society for Promoting Agriculture whose grant supported, in part, the publication of this book.

Library of Congress Catalog Card Number: 77–3810
Trade ISBN 0–06–020165–7
Harpercrest ISBN 0–06–020166–5

FOREWORD

This book is about farms, but it is not a how-to-do-it guide to farming. It is a book to be carried on trips into the countryside, particularly in the Northeast. It may help explain some of the activities and objects you see along the way. At other times the book may be browsed through like a trip to a farm itself.

Farms have changed in many ways during our nation's history. In this book we look at both the older, general farms and the modern, single-purpose farms. In every period of our history the people of the cities and towns have depended heavily upon the farmers for their food and much of their clothing. This book, like Drumlin Farm, the educational farm that inspired it, was designed to help people understand that all plants and animals, including man, are interconnected with each other and with the sun, air, rain and other features of the environment.

Farms are more than a farm family and its domestic plants and livestock; they include the land and wild plants and animals as well. All of these are found between these covers. It is our sincerest hope that this simple book will help you better see and understand the countryside that supports us all.

C.E.R.
R.J.F.
R.B.

TABLE OF CONTENTS

THE CHANGING FARM SCENE

In 1700 much of the eastern part of our country was heavily forested. The early settlers cut the trees and tilled the soil to create the farms that were to feed a new nation.

At first most of the farms provided mainly for the families that farmed them. In time the farms were enlarged and more and more livestock and crops were grown for sale to nearby cities and towns. Some of the crops were even exported.

By the mid-1800's much of the forest land in the East had been cleared and the human population was growing rapidly. In the Northeast some of the farms were abandoned as people moved into cities during the 1900's or moved west to richer farm lands.

Today many of the older farms have been blended together to make them more economical. In the place of small family farms we are finding huge agri-businesses. Many other farms have been buried by houses, cities, and highways. Some smaller family farms still remain along with evidence of others only recently lost. This book is primarily about these farms and is intended to help you gain some insight into the farm scene as you travel about.

1700

1750

1850

1920

TODOAY

7

SOIL—BASIS OF ALL THE FARM'S RICHES

SOIL is a mixture of

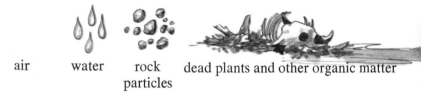

air   water   rock   dead plants and other organic matter
                          particles

The rock particles may be large or small. Particles this size are called:

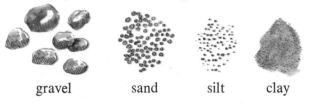

gravel   sand   silt   clay

The many possibilities of mixtures of the materials help form the many different types of soil.

Different plants prefer different types of soil. Some soils are better for trees than crops, or grass than trees. A farm has to grow the crops best suited to its soils or change the soils to conditions the desired plants prefer.

Different kinds of soil forms on:

hills   plains   wetlands

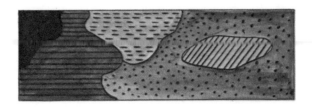

Several different kinds of soil may have formed on any farm, depending upon where it is located. Each kind of soil has different characteristics. Soil scientists can map the soil for the farmer and help him decide what crops are best for each kind of soil.

THE FARMER AND
THE WEATHER

Farmers' fortunes often hinge on the weather. Ability to predict bad weather may help them protect their crops. This may mean baling cured hay before rain ruins it; covering delicate seedlings if frost threatens; or setting up irrigation equipment to forestall drought.

"A time to sow and a time to reap"
(Ecclesiastes)

Hailstorms cause great damage and flooding can uproot plants or cause them to rot.

Long dry spells kill many crops and may speed soil erosion by wind.

Farmers have to keep track of seasonal weather patterns that govern planting and harvesting time.

FARMERS' WEATHERWISE SAYINGS:

Even with modern weather reports, local conditions vary enough that good farmers are still sharp weather observers.

Swallows flying way up high
Mean there's no rain in the sky.
Swallows flying near the ground
Mean a storm will come around.

When the dew is on the grass
Rain will never come to pass.
When grass is dry at morning light
Look for rain before the night.

READYING THE GARDEN

Spreading manure or fertilizer—
returns nutrients crops have
removed from the soil.

Plowing—mixes up the parts of
the soil and loosens it so roots can
grow easily.

Harrowing—smooths the surface
so that there is a good bed for
the seeds.

Planting—putting seeds in the ground properly
so that they will sprout and grow.

Before farmers can plant gardens or fields, the soil must be properly prepared to give the crops the best environment for sprouting and growth.

Nutrients removed by previous crops must be replaced. The soil surface must be loosened to make it easier for water and air to get to the plant roots. This also makes it easier for the roots to move through the soil.

Conditioning the soil is one of the farmer's real arts. Viewpoints vary on the best way to do it. Some prefer turning the soil over with a plow; others only chop it up with a disk or rototiller. Some use only "natural" fertilizers like animal manure and decayed plants, while others rely on "artificial" inorganic chemical fertilizers. Many use a mixture of such procedures.

PLOWING THROUGH HISTORY

Originally soil was broken with a hardened stick. Eventually a wooden plow was invented. This was later tipped with iron. In time many plows were linked together. An even later invention was the disc plow.

VEGETABLES COME FROM MANY DIFFERENT PLANT PARTS

Vegetables are an important part of most people's diet. Over the centuries man has learned to use many different parts of plants to give him food.

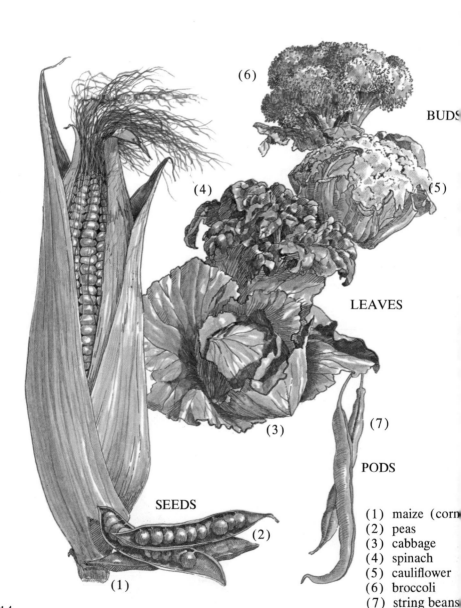

(6)

BUDS

(4)

(5)

LEAVES

(3)

(7)

PODS

SEEDS

(2)

(1) maize (corn)
(2) peas
(3) cabbage
(4) spinach
(5) cauliflower
(6) broccoli
(7) string beans

(1)

FRUIT

(9)

(8)

(11)

(10)

(12)

ROOT

STEMS

(13)

TUBERS

(14)

(15)

(8) cucumber
(9) tomato
(10) pepper
(11) squash
(12) beet
(13) carrot
(14) asparagus
(15) potato

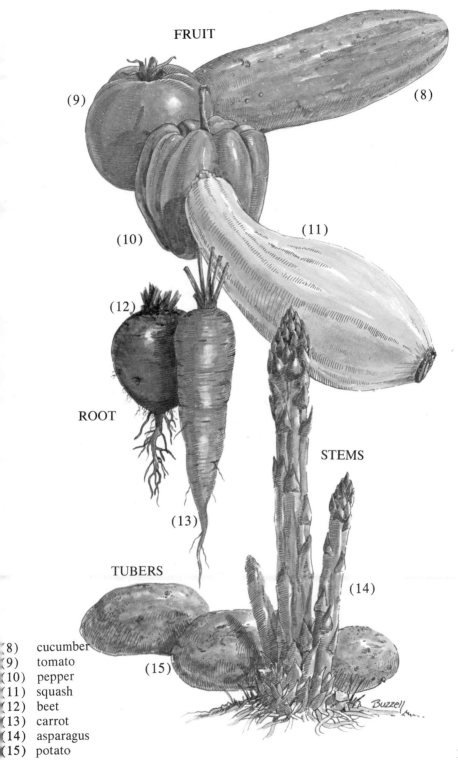

15

(1) turnip (2) parsnip (3) radish (4) carrot (5) beet (6) onion (7) potat

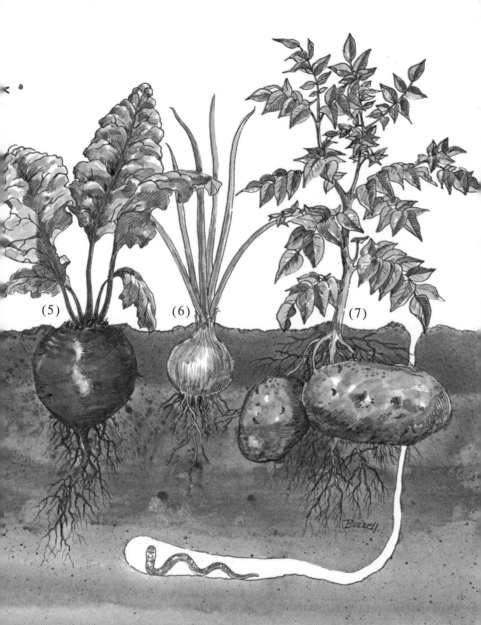

(5)   (6)   (7)

Buzzell

FOOD FROM BELOW THE GROUND

The leaves of a plant are able to turn carbon dioxide from the air and water into sugars. The plant turns these into starches and stores them in the roots of the plant.

Some plants store more such starches in their roots than others. It is these that have become our major root crops. They are particularly useful because they keep better without preservatives than most other vegetables.

17

SQUASH FAMILY

Pumpkins and squash originated in America and their edible fruits come in a wide variety of shapes and sizes. The so-called summer squashes are eaten while young and fairly soft. Winter squashes develop hard rinds and can be stored in dry, non-freezing areas for many months.

Pumpkins are used as cattle feed as well as human food.

(1)

(2)

(3)

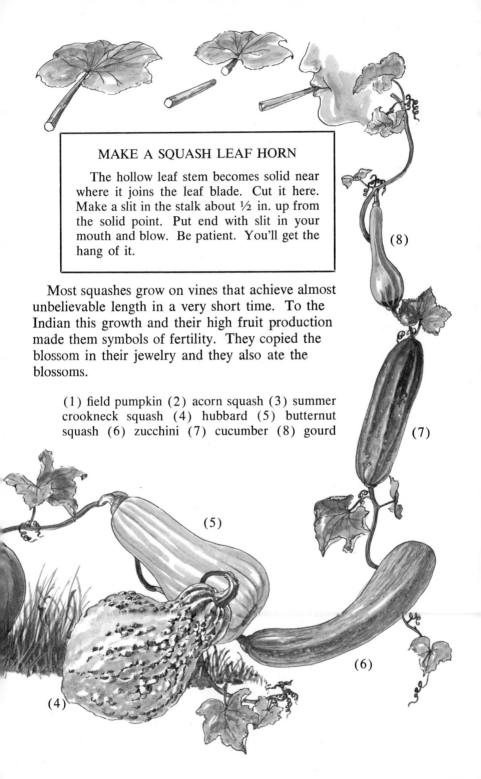

(8)

Most squashes grow on vines that achieve almost unbelievable length in a very short time. To the Indian this growth and their high fruit production made them symbols of fertility. They copied the blossom in their jewelry and they also ate the blossoms.

(1) field pumpkin (2) acorn squash (3) summer crookneck squash (4) hubbard (5) butternut squash (6) zucchini (7) cucumber (8) gourd

(7)

(5)

(4)

(6)

(1)

(1) apple blossom
 and fruit
(2) strawberry
 blossom and fruit
(3) pear blossom
 and fruit
(4) blackberry

FRUITS AND BERRIES

Gathering fruits and berries from field and roadside has long been a delightful summer and fall pastime, although today there are numbers of farms in some parts of the country that specialize in orchards of apples, peaches, pears, plums and cherries, or the raising of berry crops such as blackberries, strawberries, gooseberries and blueberries. With the major exception of the blue-

(2)

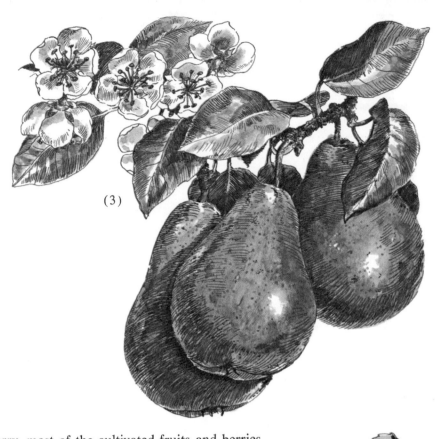

(3)

berry, most of the cultivated fruits and berries of northern sections are members of the rose family with its basically five-petaled flowers.

The rose family surrounds its seeds with a pulp that will eventually decay and form a pocket of fertilizer for its sprouting seeds; that is unless man or some other animal eats the tasty, nutritious fruit first.

(4)

(1) old fashioned hollow log hive
(2) coiled straw bee skep (hive)
(3) comb frame to fit inside modern hive

(1)

(2)

(3)

ONEYBEES

Honeybees are raised all over the world to pollinate certain
ps and produce honey. For a worker bee to make a pound of
ney, it may take as many as 40,000 to 80,000 trips from a
e to a mile and a half. The worker bees collect nectar from
wers, and through evaporation of the nectar's moisture and
nbination with the bee enzymes, the nectar is changed into
ney which is pumped into the honeycomb for storage. The
or and flavor of the honey depend upon the flowers from which
nectar has been gathered.

Bees secrete a liquid that hardens into wax scales from which
insects mold the wax comb, or cells, into which the eggs are
l or honey stored. In the complex bee society, only the queen
l lay eggs, but the thousands of workers gather the honey and
e for the young. A few male or drone bees are tolerated in
hive to fertilize the eggs.

worker

queen

drone

(1) modern hive
(2) smoker to calm bees
(3) glove
(4) head net

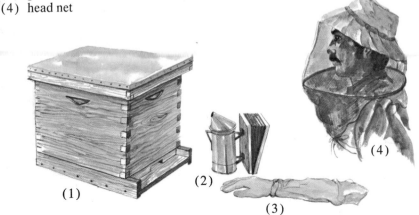

(1)

(2)

(3)

(4)

PATTERNS OF TILLAGE

Farmers must constantly guard against erosion of the soil by wind and water. Also some crops drain the soil of nutrients more rapidly than others.

Farmers frequently alternate strips of a heavily cultivated, soil-mining crop, like corn, with strips of a spreading, soil-enriching crop, like clover. This pattern is called strip cropping.

A wise farmer will plow his furrows to follow the contours of a hill thus creating a series of little dams that slow water down to soak in and not erode. This is the practice of contour plowing.

Crops may be grown on a field for several years and then the field is put into a cover crop such as clover, alfalfa, or rye grass and allowed to rest or go fallow. The following year the cover crop is plowed under as a green manure and the field replanted to a cash crop. This is the process of fallowing.

Each of these practices alone or in combination creates interesting patterns on the farmscape.

MAIZE OR CORN

popcorn

flint corn

dent corn

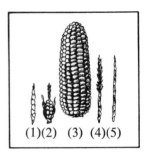

(1)(2) (3) (4)(5)

One of the great-
est cultural gifts of
the Native American
to the western civili-
zation was the tall
coarse grass called
maize. Since the white man had no word for it, he
used the general word for grain—corn. And corn it
has remained.

Maize is not found in the wild and cannot
survive today without human cultivation. It
probably originated as a cross of several near
relatives of Tripsacum grasses.

(1) hybrid maize, tripsacum, and teosinte
(2) primitive maize
(3) modern maize
(4) tripsacum—an ancestor?
(5) teosinte—another ancestor? Despite
years of research the arguments over
what is corn's ancestry continue un-
resolved.

On the right is primitive pod-popcorn. The plants to the
left illustrate various developments due to human selection.
On the far right is Corn Belt dent corn with its single stalk
and single large ear. Today there are 300 races of corn.

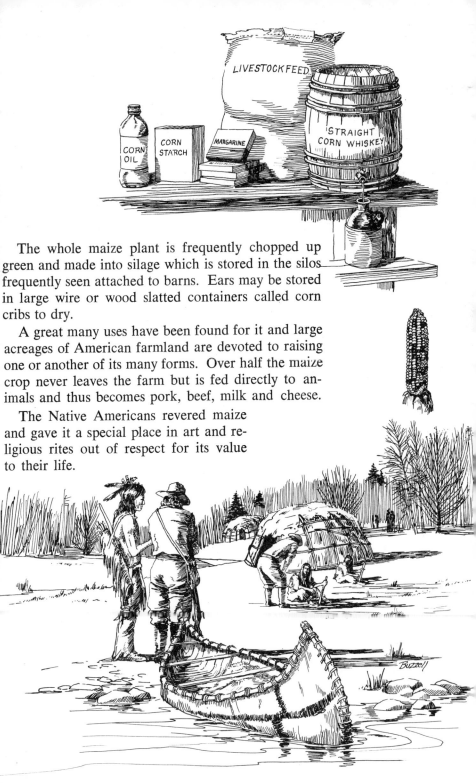

The whole maize plant is frequently chopped up green and made into silage which is stored in the silos frequently seen attached to barns. Ears may be stored in large wire or wood slatted containers called corn cribs to dry.

A great many uses have been found for it and large acreages of American farmland are devoted to raising one or another of its many forms. Over half the maize crop never leaves the farm but is fed directly to animals and thus becomes pork, beef, milk and cheese.

The Native Americans revered maize and gave it a special place in art and religious rites out of respect for its value to their life.

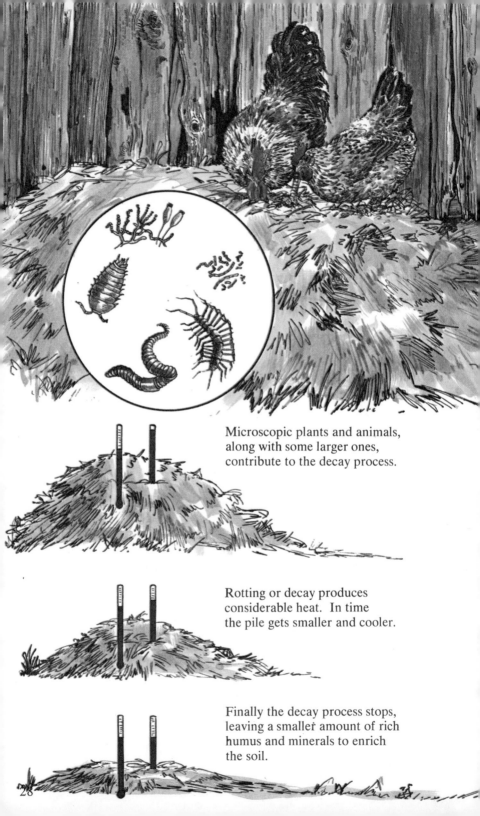

Microscopic plants and animals, along with some larger ones, contribute to the decay process.

Rotting or decay produces considerable heat. In time the pile gets smaller and cooler.

Finally the decay process stops, leaving a smaller amount of rich humus and minerals to enrich the soil.

MANURE PILE—
RECYCLING ON THE FARM

The supply of plant food elements in the soil is not inexhaustible. Like a bank account it becomes depleted if the amount withdrawn is greater than the amount deposited. With the removal of each crop, the soil surrenders some of its fertility. If an equal amount of fertility is returned by man, productiveness is maintained.

Animals eat the crops and there is a leftover waste—manure. A great deal is generated. For each thousand pounds of animal in a year sheep produce 7½ tons of manure, steers 8½ tons, horses 10 tons, cows 15 tons and hogs 18 tons.

Manure and bedding can be piled to decompose and release nutrients that can be returned to the soil for new crops to use. In the process of decomposition manure releases heat and methane gas. Heat from decaying manure can be used to warm seed beds and some farmers are learning to use the methane for fuel.

Because of the use of feedlots far from fields, tons of manure are wasted each year and are now a major pollution problem.

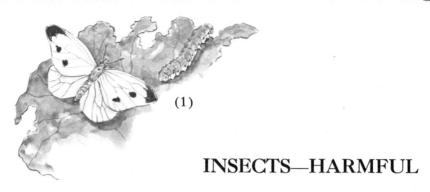

(1)

INSECTS—HARMFUL

There are more species of insects than of all other plants and animals combined. Over 88,000 insect species live in North America. Only a relatively few species are directly harmful to human interests either through spreading disease or eating crops or foods grown for human use. The farmer maintains a running battle with the harmful insects, attempting to get as much as possible of his crops and livestock for human use rather than insect use.

(1) cabbage butterfly and caterpillar
(2) colorado potato beetle and larva
(3) leafhopper and its spittlebug larva
(4) aphids—young and adults

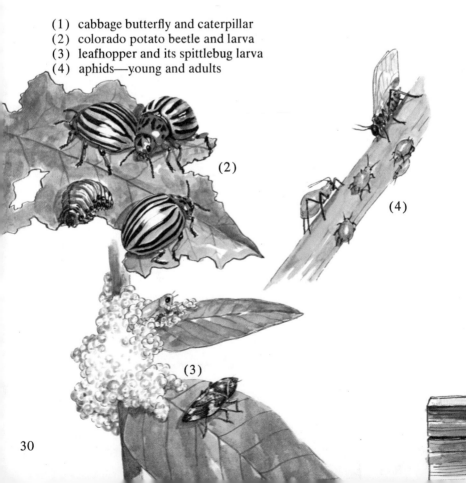

(2)

(4)

(3)

(5)

AND HELPFUL

The farmer welcomes, and labels helpful, those insect species that one way or another destroy the insects that compete with him. The farmer is also highly dependent upon those bee and fly species that pollinate his crops, for without their assistance his yield would be much reduced or even eliminated.

(5) ladybird beetle and larva feeding on aphids
(6) praying mantis feeding on grasshopper
(7) ichneumon wasp parasitizing a tomato horn worm (larva of a sphinx moth)

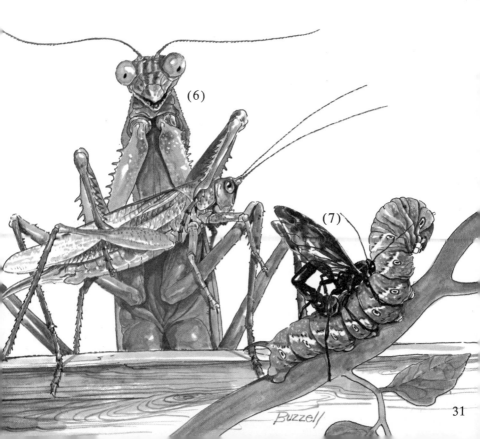

(6)

(7)

Buzzell

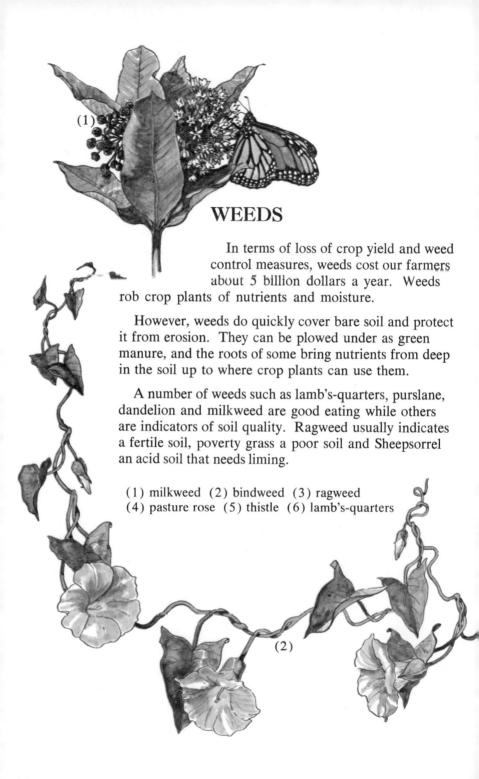

WEEDS

In terms of loss of crop yield and weed control measures, weeds cost our farmers about 5 billion dollars a year. Weeds rob crop plants of nutrients and moisture.

However, weeds do quickly cover bare soil and protect it from erosion. They can be plowed under as green manure, and the roots of some bring nutrients from deep in the soil up to where crop plants can use them.

A number of weeds such as lamb's-quarters, purslane, dandelion and milkweed are good eating while others are indicators of soil quality. Ragweed usually indicates a fertile soil, poverty grass a poor soil and Sheepsorrel an acid soil that needs liming.

(1) milkweed (2) bindweed (3) ragweed
(4) pasture rose (5) thistle (6) lamb's-quarters

(4)

(3)

(5)

(6)

Buzzell

NATURAL WEED EATERS
AND DEBUGGERS

During the nesting season most of our smaller birds feed heavily on insects. They are the primary diet for baby birds. There are also many insects whose primary diet is other insects. Where birds and predatory insects and spiders are encouraged, pest insects normally do a minor amount of damage. Chemical pesticides often destroy these natural debuggers more effectively than the pests they are designed for.

A great many songbirds feed heavily on weed seeds through fall, winter, and spring, and help to control the spread of these plants.

(1) house wren   (5) garden spider
(2) meadowlark   (6) brown thrasher
(3) ladybird beetle   (7) indigo bunting
(4) praying mantis

(7)

(6)

(5)

(4)

35

WHEAT

One of the oldest domesticated grains, wheat is known to have been grown in China at least 3,000 years B.C. Typically a crop of cool climates with less than 30″ rainfall, wheat is the "staff of life" for temperate zone peoples of five continents. It provides an excellent source of minerals and carbohydrates; is a good source of B vitamins and a fair source of vitamin A, protein, calcium and phosphorus.

American wheat fields are generally very large and are usually machine harvested.

SOME TYPES OF WHEAT

(1)   (2)   (3)   (4)

(1) Hard Red Spring—a bread wheat used for quality flour.
(2) Soft Red Winter—high starch, low protein, giving an elastic dough for pastry products.
(3) Durum—moderately starchy, high in gluten, much used for macaroni products.
(4) White—unpigmented, starchy, poor in protein and gluten, used for pastries and breakfast foods.

OATS

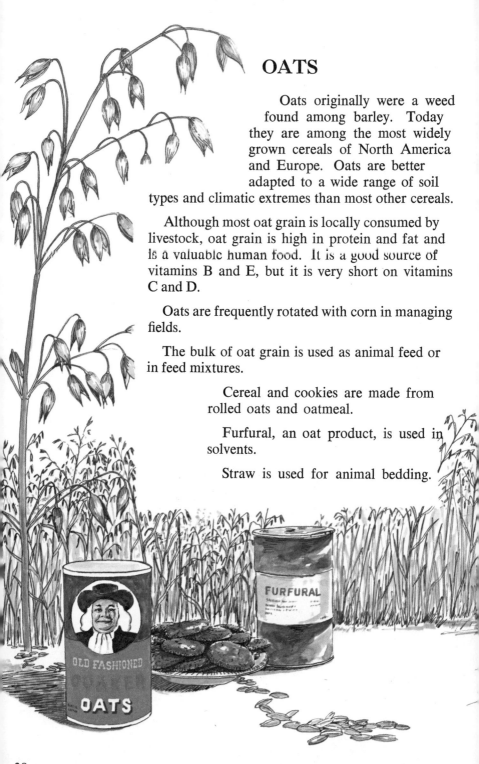

Oats originally were a weed found among barley. Today they are among the most widely grown cereals of North America and Europe. Oats are better adapted to a wide range of soil types and climatic extremes than most other cereals.

Although most oat grain is locally consumed by livestock, oat grain is high in protein and fat and is a valuable human food. It is a good source of vitamins B and E, but it is very short on vitamins C and D.

Oats are frequently rotated with corn in managing fields.

The bulk of oat grain is used as animal feed or in feed mixtures.

Cereal and cookies are made from rolled oats and oatmeal.

Furfural, an oat product, is used in solvents.

Straw is used for animal bedding.

OLD FASHIONED OATS

FURFURAL

TOBACCO

Tobacco was one of the first American crops taken to Europe. Smoking and chewing tobacco quickly became a basic habit of western man.

Although the Surgeon General's office has issued repeated warnings about the health hazard of smoking, the consumption of tobacco products continues to increase.

The leaves are gathered and dried. The best quality are used as outer wrappers for cigars. The others go into cigarette and pipe tobacco preparation.

The alkaloid nicotine from the tobacco plant yields one of the three most important insecticides known from plant sources.

Tobacco seeds are minute and must be started in flat boxes. Young plants are then transplanted to the fields. In some parts of the country, wrapper leaf tobacco is grown under the shade of cheesecloth. You may see acres and acres of cloth-covered fields shading the crop. Nearby are usually found curing barns where the tobacco leaves are hung to dry and age before being used.

normal leaf

leaf exposed
to air pollution

AIR POLLUTION
INDICATOR

It is ironic that a plant whose own products cause much air pollution is itself a sensitive indicator of many other air pollutants.

PASTURAGE

Most of our large domestic farm animals are grazers. Thus considerable farm acreage is devoted to pastures where livestock can forage for grasses and legumes such as blue grass and clover. Livestock may also be put into grain fields to graze at appropriate times.

TURN OF THE CENTURY HAYING

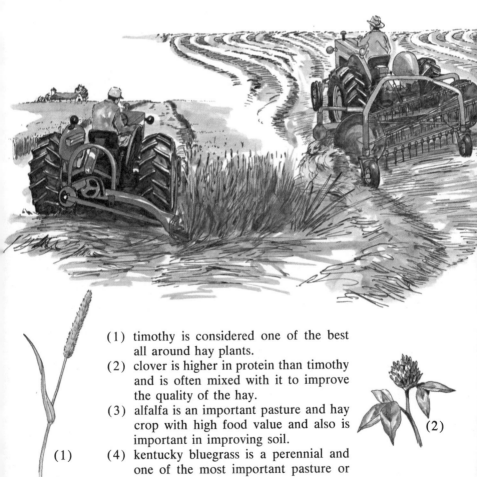

(1) timothy is considered one of the best all around hay plants.

(2) clover is higher in protein than timothy and is often mixed with it to improve the quality of the hay.

(3) alfalfa is an important pasture and hay crop with high food value and also is important in improving soil.

(4) kentucky bluegrass is a perennial and one of the most important pasture or forage plants.

HAYING TODAY

AND HAY

Grazers must be fed in winter also, so fields of alfalfa, timothy, clover, and other grasses and legumes are grown. At the peak of their food value they are cut, dried and stored for winter use. Hay is usually cut twice a year although in some regions as many as eight cuttings are made.

(3)

(4)

43

(1)

(6)

(2)

(3)

(4)

44

ECOLOGY OF THE FIELD

Ecology deals with interrelationships. The sun's energy is trapped by the green plants for their growth and maintenance. Plant-stored energy is taken by plant-eating creatures like the rabbit, woodchuck, mouse, and grasshopper for their own use. Seeking energy, the bobolink may take it from plant seeds or plant-eating insects. Meat eaters, like the fox, hawk and snake, secure their energy from that gained by the plant-eaters or smaller meat-eaters that they catch and eat.

When the plants and animals die, their bodies break down, releasing the materials to soil and air. From these materials new plants and animals will be made and the flow of energy and the cycling of materials will go on. The raw materials in the field may also be useful to young woodland plants. If grazing, mowing, or fire don't kill the woodland plants they may grow up and crowd out the field plants. In time the field plants and animals may be crowded out by those of the forest.

(1) kestrel (2) woolly bear caterpillar (3) orb weaver spider with grasshopper (4) leafhopper (spittlebug) (5) milk snake (6) flicker (7) cottontail rabbit (8) woodchuck (9) red fox

FLYING MOUSETRAPS— NIGHT AND DAY SHIFTS

Rats, mice, ground squirrels and other rodents eat a great deal of grain and crops that might otherwise be used by man. Traps and poisons are of some value but they present hazards as well. Fortunately there are two groups of winged predators whose primary food is these rodents— they are the hawks and owls.

Owls are the night shift of the rodent patrol and though less often seen are no less effective. Indeed, many rodents are most active at night.

Hawks are daytime birds and thus form the day shift of the rodent patrol. Farmers used to believe the hawks were a threat to their poultry and shot the birds. Science has proved otherwise and hawks and owls are now generally protected by law.

Unfortunately both hawks and owls are less abundant today due to mistakes of the past with hunting laws and pesticides. Some species are threatened with extinction.

Buzzell

GAME BIRDS

Two of America's
most popular game
birds are birds of our
farmlands.

The pheasant was introduced to this country
and found a home here particularly on
farmlands. It is our most abundant game bird.

The bobwhite
quail is a native and thrives
wherever there are brushy
fencerows. It is valuable as an
insect eater on the farm as well as
for the hunting sport. Bobwhite have
been depleted in many areas of former
abundance due to use of pesticides on crops.

Farm woodlots may harbor the ruffed grouse.

The wild turkey is making a comeback in several states and may be seen in some wilder woodlots.

—S. Buzzell

BUCKWHEAT

In 1866 the United States produced some 22 million bushels of buckwheat, but by 1969 we were producing only ¾ of a million bushels. Buckwheat does not lend itself to the mechanization of today's agri-business. It is essentially a small family farm crop.

Buckwheat is used primarily for chicken feed and flour. Buckwheat flour pancakes are especially delicious. As the interest in natural, or "health" foods rises we may again see an increase in the amount of this crop produced.

Buckwheat nectar makes a very fine honey; the seeds are excellent wildlife food, particularly for pheasant and quail; and the plant itself is a first-rate green manure for enriching the soil.

SOYBEANS

The soybean, a native of China, was introduced into the United States in the 1930's. Since then we have become the world's major producer of this crop. Indeed, it has become one of our top three cash crops.

The value of soybeans lies in their protein and fat content and the fact that as members of the clover family their roots contain nitrogen-fixing bacteria which help feed the plant and enrich the soil. Soybeans were originally grown in this country primarily to feed livestock, but today they are used in a wide variety of ways for industrial use and human food. Artificial meat products, shortenings, margarine, breakfast foods and soybean oil are among the products made. Approximately 170 grams of soybean concentrate will supply the daily requirements of proteins, vitamins, and minerals for the average adult human.

As man finds more uses for this plant, he will be using more of it directly and less will be fed to livestock. Already farmers are finding protein-rich soy feed grains soaring in price, and the hungry people of the world want to eat the soybeans in vegetable form.

red jungle fowl   mallard duck   muscovy duck

ORIGINS OF DOMESTIC ANIMALS

People began keeping tamed animals as sources of meat, hides, and probably milk shortly after the end of the Ice Age—about ten to twelve thousand years ago. Through careful selection of breeding stock, herdsmen and farmers across the centuries have changed the initial wild stock into the breeds of domestic animals we know today.

From the Mongolian wild horse and the European tarpan were developed both today's riding and draft horses. The now extinct aurochs was the ancestor of most of our dairy and beef cattle breeds. The wild boar has been changed into the many varieties of domestic swine.

Most of our common breeds of sheep derive from the handsome mouflon. Their close relatives the dairy goats were developed largely from the Cretan wild goats.

Birds were not ignored. The red jungle fowl, first domesticated for sport fighting, was transformed into our many breeds of chicken. The common mallard gave rise to most of our domestic ducks, the main exception being the muscovy duck of South America. The Americas also gave us the wild ancestor of the domestic turkey.

aurochs

wild boar

Cretan wild goat

mouflon

tarpan

SHEEP AND GOATS

Although Angora goats are raised for their wool, most goats in this country are dairy breeds and are a source of milk, fine leather, and a meat called chevon. Goats' milk is naturally homogenized, has more fat, and less lactose but the same water and protein content as cows' milk. Many people who cannot digest cows' milk can drink goats' milk. Goats are more curious and intelligent than sheep.

(1) saanen buck
(2) saanen doe
(3) toggenburg doe

(1)

(3)

(2)

(4)

(5)

(4) french alpine doe
and kid
(5) nubian doe

Sheep and goats are closely related animals. Sheep are raised primarily for wool and the meats lamb and mutton. The various breeds have different qualities of wool and meat and differing abilities to thrive in a variety of pasture conditions.

Sheep are primarily grazers, while goats are naturally browsers. Both must be carefully managed so that they do not destroy the plants of their pastures.

(1) suffolk ram
(2) cheviot flock

(3) southdown ewe
(4) hampshire ewe
(5) merino ram

Dogs are often trained to help the shepherd herd the sheep.

SHEARING AND SPINNING

Wool is a specialized form of hair. Each hair, or fibre, has tiny scales that let the fibres cling together. It grows in tufts of up to a dozen fibres. Glands at the base of the wool give the wool a greasy coating.

Using hand or electric clippers, men shear the wool from the sheep. A good shearer takes the fleece of wool off the sheep in one piece. The raw fleece is then bundled and sent off to have the grease removed. The grease itself is used to make lanolin used in skin softeners.

The clean wool is then carded, that is, run between special combs to get all the fibres running in the same direction. The resulting batch of fibres is called a rollag.

The fibres of the rollag are then twisted tightly and the little scales of the wool hold the twist. Many ways have been devised to twist the fibres. The simplest perhaps is the spinning drop spindle but the most familiar is the spinning wheel. Actually the spinning wheel is only a drop spindle turned sideways and turned with a wheel. Of course today giant machines do the job.

POULTRY

Chickens are poultry, but so too are ducks, geese, turkeys, and guinea fowl. Poultry is a word that includes a variety of fowl kept for the production of meat and eggs.

On the American market, chickens and turkeys are the most important meat birds with ducks a distant third. Geese and guinea fowl are specialty items generally only locally available.

Chicken eggs dominate the market. Although duck eggs can be found in some supermarkets, they are more likely to be found at roadside farm stands.

When they reach suitable weight, 3½ to 4 lbs. for broiler/fryers or 5½ to 6 lbs. for roasting chickens, they are crated up and delivered to the butchers. There they are killed, plucked, chilled or frozen, and packaged for marketing.

The developing birds are put into buildings or out onto ranges to eat and put on weight.

Baby chicks are at first kept warm under heated brooders.

21 days

18 days

JOURNEY OF AN EGG

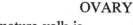

OVARY

A mature yolk is released by the ovary into the funnel.

FUNNEL

The yolk takes about 15 minutes to travel through the funnel into the magnum.

MAGNUM

During the two hours and 45 minutes the yolk is passing through the magnum it gets several layers of albumen or egg white.

ISTHMUS

The inner and outer shell membranes are added to the developing package by glands in the isthmus. It takes about 1 hour and 15 minutes to pass through the isthmus.

UTERUS

The eggshell is secreted around the nearly complete egg by the uterus. The egg remains in this organ about 20 hours and 45 minutes.

VAGINA

Muscles in the vagina force out the finished egg about 25 hours after the yolk was first released.

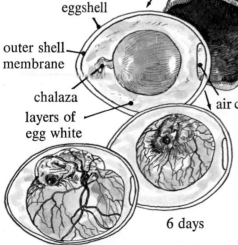

eggshell

outer shell membrane

chalaza

layers of egg white

air cell

6 days

12 days

About ½ hour after one egg is laid, another yolk may enter the funnel. Thus a good producing hen may lay about one egg a day. Some hens lay over 300 eggs a year.

63

BEEF CATTLE

(1)

Beef cattle have been bred to produce the greatest amount of meat in the shortest period of time.

Since the most valuable cuts in the beef animal are taken from the loin, ribs, and thighs, these parts of the animals have been emphasized through selection.

As with dairy cattle, most of our breeds were developed in Europe. Such northern cattle fare poorly in hot dry climates. To remedy this, the Zebu cattle from India have been crossed with European breeds to create new heat-hardy, insect-resistant ones.

Male beef calves, not to be kept as breeding bulls, have their reproductive organs removed. They are then known as steers. Steers are the major meat animals. Most young cows are kept to produce more steers.

(1) *Aberdeen Angus.* Though small, they produce carcasses of very high quality.

(2) *Hereford.* Withstands heat, drought and winter exposure. Lacks Angus quality.

(2)

(3) *Charolais.* A large rugged breed. Does well in hot weather and reasonable cold.

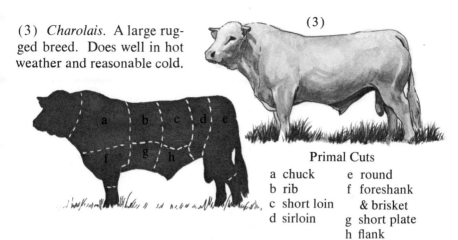

(3)

Primal Cuts

a chuck	e round
b rib	f foreshank
c short loin	& brisket
d sirloin	g short plate
	h flank

Beef cattle are generally grazed on the range and pasture while young. When they reach a proper weight they are brought to feed lots where they are fattened on grain. They are then shipped to slaughter houses where they are butchered. The meat goes to the markets and the hides are made into leather. Horns and hooves are used for a wide variety of products such as glue, while excess fat is made into such things as women's cosmetics.

(4) *American Brahman.* Developed in America from several breeds from India. More able to tolerate high temperatures than other breeds. Insect resistant.

(4)

Buzzell

DAIRY CATTLE

Most modern breeds of dairy cattle were developed in Northern Europe. This was probably because only in northern climates was it possible to make butter and cheeses which would preserve the excess milk supply from spoilage.

Female, or heifer, calves were kept for the milk they would eventually produce. Bull calves not to be kept for breeding were either butchered or altered to create oxen. Oxen were used as draft animals noted for their great strength and docile disposition.

After a cow has had a calf, milk is produced by glands located in her udder. The milk drains to the teat. From here it can be squeezed out by the calf, human hands, or special milking machines.

Holstein-Friesian. Predominant and largest dairy breed except in the South. Milk is lowest in butterfat, ave. 3.7%, and is whiter than that of other breeds.

Jersey. A small, early-maturing breed. Does not produce quantity most commercial dairies require. Butterfat content ave. 5.3%.

Guernsey. Average in size, temperament, constitution and early maturing. Guernseys rank 1st or 2nd in number in 42 states. The milk is golden color with 4 to 6% butterfat.

Ayrshire. A medium-sized, rather excitable breed. They are unsurpassed in foraging ability. They produce an excellent flow of milk and ave. 4% butterfat.

Holstein-Friesian

Jersey

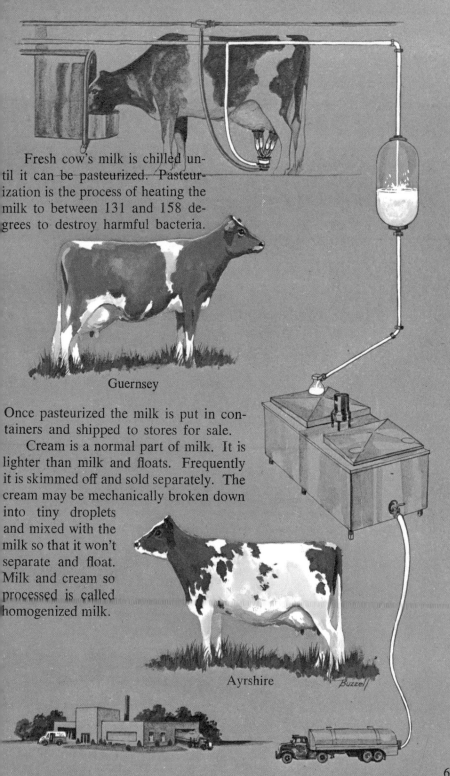

Fresh cow's milk is chilled until it can be pasteurized. Pasteurization is the process of heating the milk to between 131 and 158 degrees to destroy harmful bacteria.

Guernsey

Once pasteurized the milk is put in containers and shipped to stores for sale.

Cream is a normal part of milk. It is lighter than milk and floats. Frequently it is skimmed off and sold separately. The cream may be mechanically broken down into tiny droplets and mixed with the milk so that it won't separate and float. Milk and cream so processed is called homogenized milk.

Ayrshire

Buzzell

HOGS

From a dollar and cent angle hogs are the most valuable farm animals. They reproduce and fatten rapidly and surplus animals always find a ready market. Hogs paid off the mortgage on many a struggling farm.

From the middle 1800's farmers selectively bred from tall, slow-growing country hogs new improved types that reproduce faster and produce quantities of muscles and fat. Two major types of hogs were developed—lard types and bacon types. *Poland-China* and *Duroc-Jersey* are common lard hogs while *Hampshire* and *Yorkshire* are bacon breeds.

In many parts of the country hogs are grazed in the fields and

fattened on grain. In the northeast hogs are often fattened on swill. Swill is a mixture of all kinds of wet garbage. Hog raisers are still the only garbage disposal system in many communities. Laws generally require the garbage to be cooked before being fed to the hogs.

Adult male hogs are called boars and the females are sows. The young are pigs. Increasingly pigs are raised entirely in easy-to-clean concrete pens. Pigs are shipped to market generally at about 7 or 8 months when they reach a weight of about 220 lbs.

(1) yorkshire
(2) poland china
(3) duroc
(4) hampshire

THE HORSE TRIBE

Horses were domesticated over 4,000 years ago. Originally they were kept for milk and meat as well as for riding and draft animals. Before the internal combustion engine, horses provided most of the transportation and much of the power for plowing and pulling on American farms.

Newborn horses are called foals. A colt is a young male while a young female is a filly. Adult females are called mares and the males are stallions. Geldings are males that have been castrated.

Horses are not usually worked until they are two years old but they may work for twenty years. Horses will live up to 35 years.

(1) belgian draft horse
(2) shetland pony
(3) quarter horse

(1)

(2)

(3)

The donkey was the poor man's beast of burden. It was less temperamental than the horse and able to get along on poorer feed. Donkeys were most valuable for mating with horses to create mules. Mules combined the strength of the horse with the intelligence and thriftiness of the donkey.

(4) mule
(5) donkey or ass
(6) sicilian donkey or burro

HORSESHOEING AND

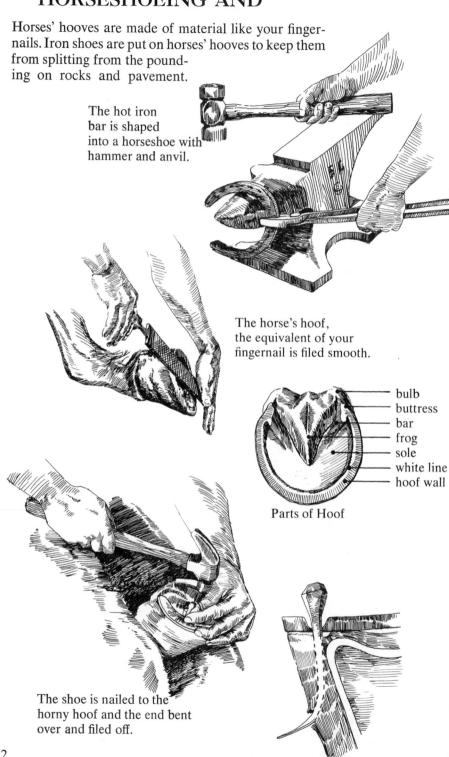

Horses' hooves are made of material like your fingernails. Iron shoes are put on horses' hooves to keep them from splitting from the pounding on rocks and pavement.

The hot iron bar is shaped into a horseshoe with hammer and anvil.

The horse's hoof, the equivalent of your fingernail is filed smooth.

- bulb
- buttress
- bar
- frog
- sole
- white line
- hoof wall

Parts of Hoof

The shoe is nailed to the horny hoof and the end bent over and filed off.

ITCHING UP THE TEAM

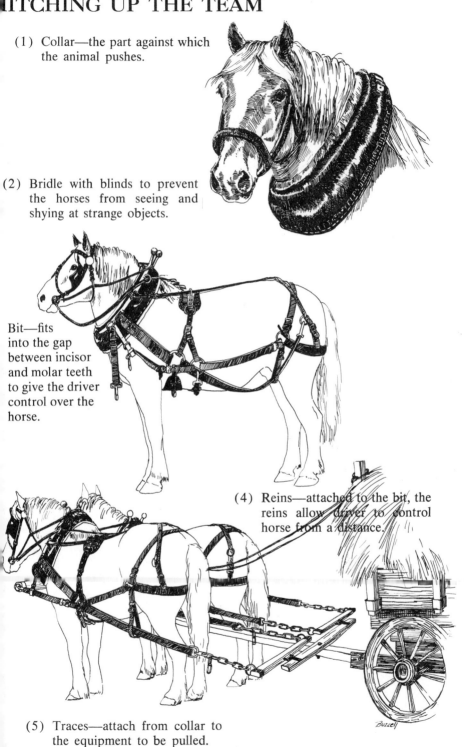

(1) Collar—the part against which the animal pushes.

(2) Bridle with blinds to prevent the horses from seeing and shying at strange objects.

Bit—fits into the gap between incisor and molar teeth to give the driver control over the horse.

(4) Reins—attached to the bit, the reins allow driver to control horse from a distance.

(5) Traces—attach from collar to the equipment to be pulled.

Buzzell

THE FARMER'S COMPETITORS AND SOMETIME HELPERS

The house mouse and rat have learned to live with man and take most of their food from him. They eat millions of dollars worth of grain every year; they foul food with their droppings, spread disease, and rats even kill and eat eggs and young poultry.

Woodchucks may raid the vegetable garden for young greens and their burrows create a hazard to livestock and machinery. However, the burrows provide homes for other wildlife. Rabbits, too, raid the gardens and meadow mice may eat root crops and ripening grain.

The fox and the skunk may take an occasional chicken but make up for it by eating the mice and woodchucks and insects. Their fur may also provide some occasional added income. Raccoons take a toll in the cornfield and among the small fruits and are smart enough to make the farmer really work to catch them but they also eat rodents, insects and weeds.

Deer may put meat on the farmer's table but they can make a shambles of a garden. Wildlife is all part of a farm, contributing to its success as a farm in some ways, detracting from it in others. It is all a matter of degree. Man shares the land with them.

(1) red fox
(2) cottontail rabbit
(3) woodchuck
(4) whitetail deer
(5) raccoon
(6) striped skunk
(7) norway rat

75

THE FARM POND

Farm ponds provide water for livestock and irrigation. They also provide a source of water to put out any fires. Many farmers stock their ponds with fish for additional food and recreation. Swimming is often an added benefit.

Even man-made farm ponds soon develop a natural plant and animal community. Microscopic plants feed tiny animals which in turn feed little fish which themselves are often food for bigger fish. Cattails, flag, water lilies and other shallow water

Buzzell

plants provide cover for many of the fish and other pond creatures like frogs, turtles, and muskrats. Many birds are also attracted to the pond environment. All the members of the farm pond share a variety of interrelationships.

Healthy farm ponds have a variety of life. Where too much manure, chemical fertilizer, or salt drains into the pond, important changes may occur that favor a few species, like algae, but make the pond unsuitable for many of the plants and animals that would normally live there. Such ponds are much less useful.

(1) redwinged blackbird (2) painted turtle (3) water lily (4) yellow perch (5) pickerel (6) mayfly (7) sunfish (8) bullhead (9) whirligig beetle (10) water strider (11) green frog (12) snapping turtle (13) dragonfly (14) raccoon (15) muskrat (16) mallard duck (17) cattail

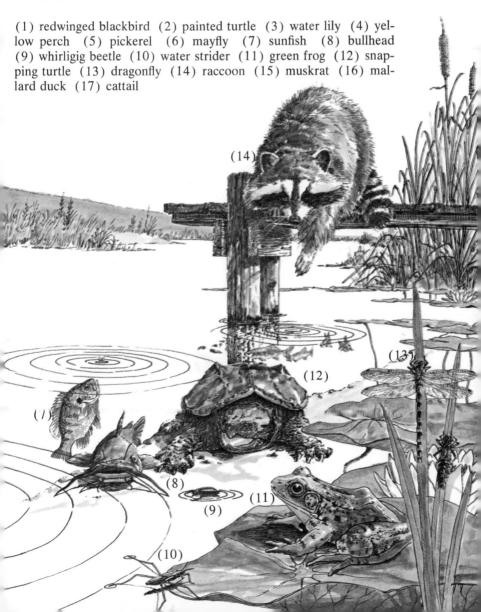

WILDFLOWERS OF WOODS AND FIELDS

The majority of woodland flowers appear early in the year, using energy stored from the year before. Before the opening leaves form a canopy that shuts off much of the essential light, the woodland plants manufacture most of their food for next year's flowering.. This is stored underground and the above-ground parts die back. Woodland flowers tend to wilt quickly and picking is likely to damage or kill the whole plant.

(1) jack-in-the-pulpit (2) skunk cabbage (3) marsh marigold
(4) bloodroot (5) pink lady's-slipper (6) painted trillium

Field flowers tend to make the food they need for flowering in the same year they flower, so field flowers are more abundant in summer and fall, although there are spring bloomers as well. Most of our cultivated flowers were developed from field flowers rather than woodland species. Field flowers generally last longer when picked and, for many species, picking actually increases the number of blooms. However, it is still wise to observe the rule of ten—that is, for every ten blossoms you can see you may pick one.

(1) aster (2) chicory (3) dandelion (4) black-eyed susan

THE FARM WOODLOT

The farm woodlot was a basic part of many small farms. In earlier days when many farm machines and tools were made in large part of wood, the woodlot produced hickory and ash for tool handles, oak for wagon wheel spokes and wagon floors, and dogwood for axles. Pine and oak were important woods for making furniture and building buildings. Fencing was made from split chestnut and white cedar. Indeed, the American forest provided a great variety of woods to meet a variety of needs.

Today other materials are used to meet many of the farmer's needs, but the woodlot is still useful. Wood is still cut for fuel either to be used on the farm or sold as a cash crop. Nuts are gathered and woodlot wildlife hunted. Some farmers are raising Christmas trees on their woodlots while others tap their maple trees. Land used for woodlots is usually land that is of poor value for crops or grazing, yet the woodlot makes a positive contribution to the aesthetics and economics of the farmscape.

(1)
(2)

THE MAPLE GROVE OR "SUGARBUSH"

Leaves of maple trees manufacture sugar which is stored in the roots over the winter. In spring the sugar-rich sap rises to help new leaves begin their work. Sugar maples produce the most, but other maples produce sugar too.

Indians learned to gather maple sap and boil it down for syrup and sugar and they taught the white man the process. When late winter or early spring temperatures reach 36°F. and night temperatures are below freezing, the sap begins to flow upwards and it is tapping time. Buckets may be individually hung to collect sap as in the old days and carried to the sugar house. Increasingly, plastic pipes are being used to carry sap directly from the trees to the evaporators where the water is boiled away. Once temperatures, night and day, stay above freezing, the maple tapping season is over.

Making maple syrup is something you can try. A hole, about ⅜ inch in diameter, is drilled slightly upward for about 1½ inches into the maple tree at waist height. The spile, either purchased, or made from sumac or elder, which has been appropriately tapered and with the pith pushed out, should be inserted snugly. From this, or a nail just above it, the sap bucket is hung.

The collected sap is boiled to evaporate away the water and concentrate the sugar. From about 40 pints of sugar maple sap the yield is about 1 pint of syrup. From other maples it will take more sap for the same yield.

(1) collecting vat
(2) sled
(3) evaporating house
(4) collecting bucket
(5) spile
(6) taphole

STONE WALLS AND OTHER FENCES

Fences are built primarily to make boundaries or keep livestock enclosed. They generally require a great deal of labor. Early fences were made from the most abundant materials on the land—wood or stone. Rocky New England is full of stone walls, often through what today are woodlands. In the midwest people tended just to make stone piles in the middle of the fields.

Today wooden fences and stone walls have largely given way to woven wire fences. Stock fencing often has more wires at the bottom than the top to keep small animals from getting in or out. Wire netting in a hexagonal pattern is generally used on poultry runs.

Wire cable was used to enclose cattle and horse pastures but it couldn't stand up to the cattle leaning on it. This gave rise to the invention of barbed wire about 1867. In 1874 Joseph Glidden created a practical machine to make barbed wire and it began to be used on a large scale. It was relatively easy and cheap to install. A number of different patterns of barbs have been used over the years.

On a smaller scale, plain wire is used to fence horse and cattle runs in heavily populated areas. A pulsating charge of electricity keeps the animals from leaning on and breaking the fence. Living fences of thorny plants like multiflora rose and osage orange are also used to contain livestock.

(6)

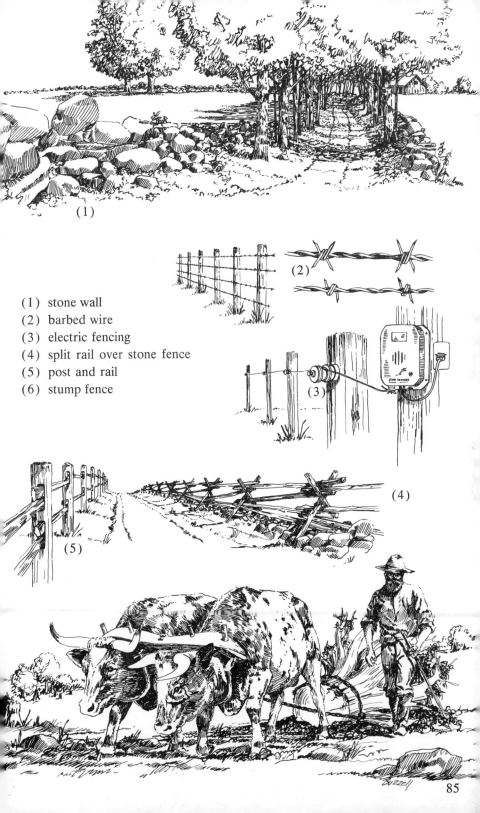

(1) stone wall
(2) barbed wire
(3) electric fencing
(4) split rail over stone fence
(5) post and rail
(6) stump fence

85

FERTILIZERS

Plants need food just as we do. They must get 15 or more kinds of food, or elements, from the surrounding air, water and soil. Twelve of these elements or nutrients come from the soil. As the plants grow, they remove these elements from the soil and make them part of the plant itself.

When the plants are harvested and removed from the field where they grew, the soil is left behind minus the elements that were removed. These elements must be replaced, or before long the soil will not be able to grow healthy plants.

The elements that are used up most quickly are nitrogen, phosphorus, potassium, calcium and magnesium. Animal and plant wastes such as manure, ground bone and dead plants are rich in these elements. For centuries on small farms, such material was mixed into the soil to replace the elements that had been removed by the past year's crop. Such materials are not sufficient for mechanized, large-scale, single crop farms.

The chemical industry came to the rescue with fertilizers that could be manufactured in large quantities and distributed easily

by machines. These commercial fertilizers seemed to be a boon to agriculture and they aided in increasing crop yields and opening up new lands for farming.

Today there are serious questions being raised about the way and the extent to which these fertilizers are being used. For one thing, they take large quantities of energy and increasingly scarce supplies of natural gas and oil in their manufacture. Carelessly used, the fertilizers quickly wash away, and fertilize ponds and streams. Such fertilized streams grow large amounts of algae and water weeds that choke the waterways and make the water unsuitable for many uses. Also, although they provide nutrients for the plants, these fertilizers often disturb the basic structure of the soil so that it holds less air and water. This makes the soil less suitable for growing plants.

Commercial fertilizers have been a typical quick-fix solution for a complex problem. They have certainly had a short-range benefit, but the future will undoubtedly produce a more sophisticated way to condition soil and replace plant nutrients.

PESTICIDES

Farmers have always resented sharing the fruits of their labors with insects, fungi, and other wildlife. For centuries their tools for fighting off such pests have been rather primitive. The pests continued to take their toll, occasionally destroying a whole crop.

The main forces against pests have been other forms of wildlife that fed upon the pests, simple mechanical devices like traps, and natural plant and mineral poisons. During World War II new chemicals were created that were extremely poisonous to insects and many other creatures. DDT was the first of these miracle pesticides. Its chemical relatives soon followed. It appeared that man now had the weapons to eliminate the pests.

Unfortunately, time has shown these chemical weapons to be two-edged swords. They killed far more than their primary targets, often killing the pests' natural enemies more effectively than the pests themselves. They killed other forms of beneficial wild-

life and induced cancer in laboratory animals and man. Since they are very long lasting, the chemicals build up in food chains from harmless quantities to harmful quantities. What's more, the more offspring a species has at a time and the more litters of offspring a year the species has, the more likely it is to produce some young that are not poisoned by the chemical. Soon the species is made up largely of individuals that are poison resistant. For slower reproducing species, it takes much longer if it can happen at all. Thus the pests tend to develop resistance to the poison while their enemies do not. The quick-fix miracle has backfired again.

Modern farmers use a combination of biological controls geared to specific pest targets, short-lived chemicals, and crop management procedures to fight insect and plant pests—and like it or not, insects, weeds, and wildlife still will get a small share of man-reared crops.

OLD AND NEW

Where once a farmer plowed a small area with a horse or an ox, today's farmer uses a 100 horsepower tractor to pull five or six steel plows at once over a large area. Farming has generally changed from a process requiring much human labor to one demanding much machinery and fossil fuels. The increased pressure for more food from less total land has promoted the development of monstrous pieces of machinery to till, plant, and harvest the crops.

The variety of modern farm machines saves much human drudge labor and lets fewer people farm more acres each. However, we pay the price of much greater amounts of energy and mineral resources to get our food. Smaller, diverse farms have largely disappeared in favor of larger one- or two-crop farms. Today's farming has become increasingly industrialized.

ENERGY ON THE FARM

In the early days the biological engine of muscle, human or animal, fueled by food provided most of the power. Mechanical engines fueled largely by electricity or fossil fuels have made farm life easier and more profitable. Many hard and tiresome jobs, such as pumping water, running the cream separator or washing machine, cleaning grain, grinding feed, mixing concrete and shelling corn, are now handled at the farmer's convenience with machinery. Naturally, energy consumption on the farm has soared.